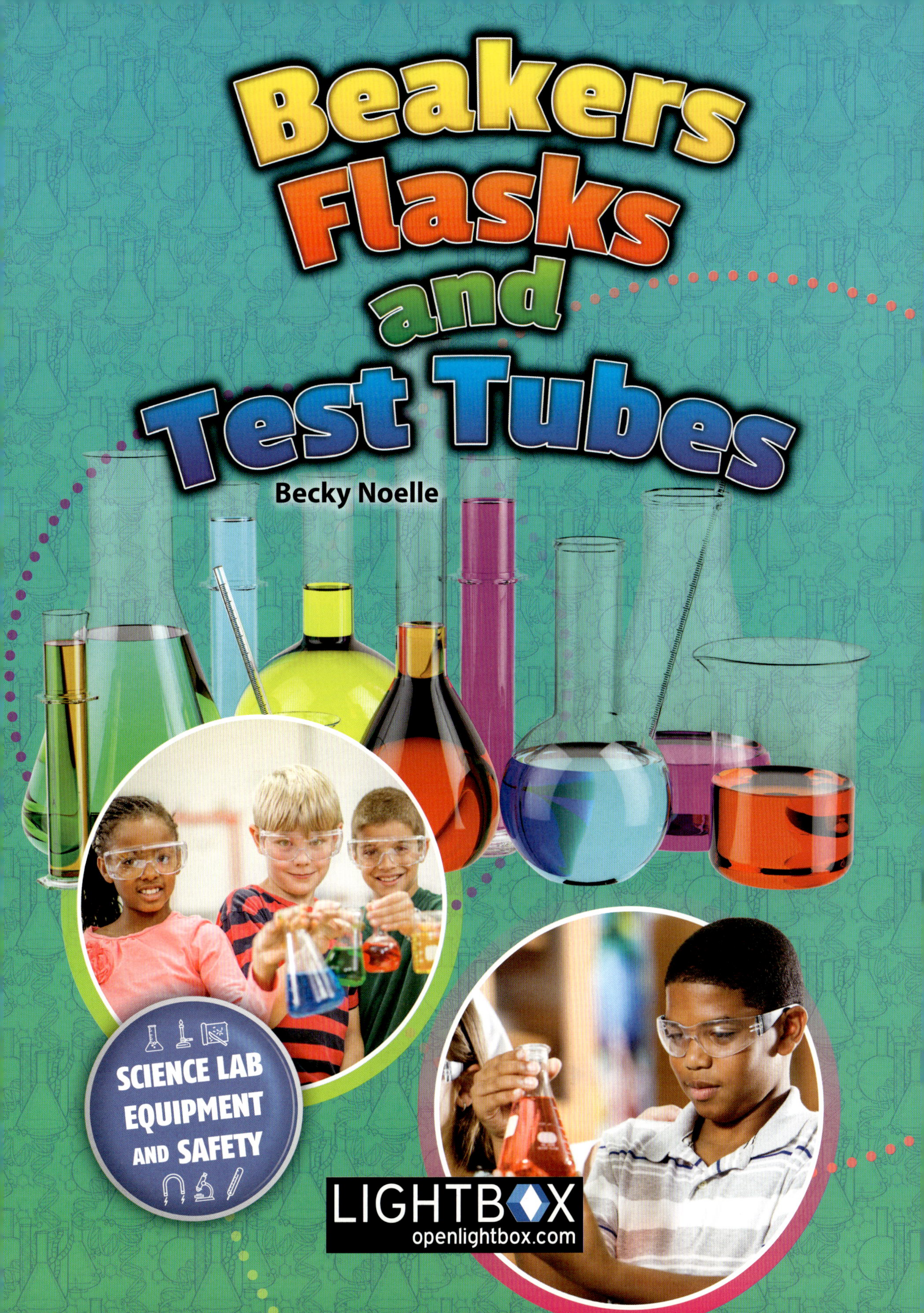
Beakers
Flasks
and
Test Tubes
Becky Noelle
SCIENCE LAB
EQUIPMENT
AND SAFETY
LIGHTBOX
openlightbox.com

Go to
www.openlightbox.com
and enter this book's
unique code.

ACCESS CODE

LBXR9938

Lightbox is an all-inclusive digital solution for the teaching and learning of curriculum topics in an original, groundbreaking way. Lightbox is based on National Curriculum Standards.

LIGHTBOX SUPPLEMENTARY RESOURCES

SHARE
Share titles within your Learning Management System (LMS) or Library Circulation System

CURRICULUM
Find national and state curriculum correlations

CITATION
Create bibliographical references following the Chicago Manual of Style

STANDARD FEATURES OF LIGHTBOX

AUDIO High-quality narration using text-to-speech system

ACTIVITIES Printable PDFs that can be emailed and graded

SLIDESHOWS Pictorial overviews of key concepts

VIDEOS Embedded high-definition video clips

WEBLINKS Curated links to external, child-safe resources

TRANSPARENCIES Step-by-step layering of maps, diagrams, charts, and timelines

INTERACTIVE MAPS Interactive maps and aerial satellite imagery

QUIZZES Ten multiple-choice questions that are automatically graded and emailed for teacher assessment

KEY WORDS Matching key concepts to their definitions

Lightbox Grades 3–5 Subscription
ISBN 978-1-5105-5424-5

Access hundreds of Lightbox titles with our digital subscription. Sign up for a **FREE** subscription trial at **www.openlightbox.com/trial**

CONTENTS

2 Lightbox Access Code
4 Mixing It Up
6 History of Beakers, Flasks, and Test Tubes
8 Types of Glassware
10 How They Work
12 Safety in the Lab
14 The Scientific Method
16 Experiment: Mystery Powder
18 Experiment: Rusty Nails
20 Activity: Make Elephant Toothpaste
22 Quiz
23 Key Words/Index
24 Log on to www.openlightbox.com

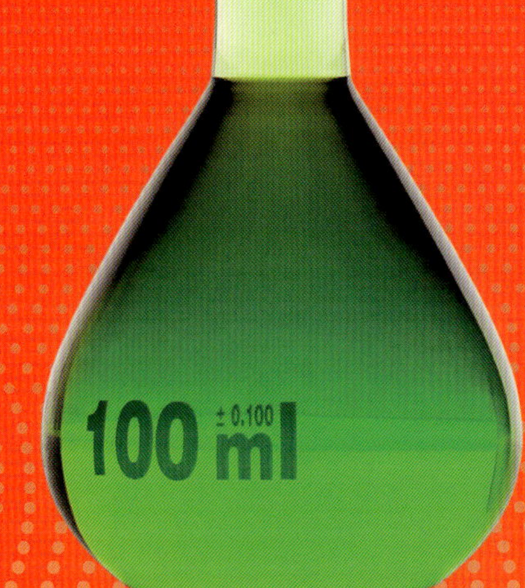

MIXING IT UP

Scientists carry out many **experiments** in science laboratories, or labs. They use special **equipment** to test their ideas and find answers to their questions. Chemistry is a type of science. Scientists called chemists often mix **chemicals** to make new products. Chemistry has its roots in a practice called alchemy. In the past, alchemists mixed different liquids and solids together. Some of them even tried to make gold or a potion that would let them live forever.

Chemists use glass containers, or glassware, in their experiments. These include beakers, flasks, test tubes, and more. Throughout history, chemists have designed glassware for different purposes. Some past chemists were also glassblowers who could make their own glassware. Today, companies make glassware in factories and sell it to scientists and science labs.

Some lab **glassware** can withstand temperatures up to **932° Fahrenheit** (500° Celsius) **without breaking**.

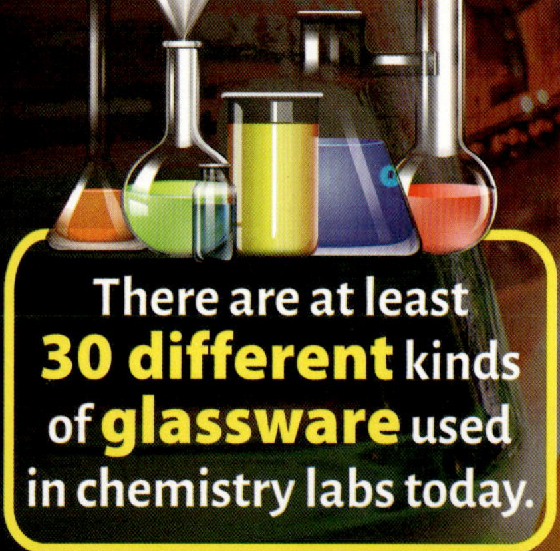

There are at least **30 different** kinds of **glassware** used in chemistry labs today.

90
80
70
60
50
900ml
700
100 ml
APPROX VOL
80
60
40
APPROX VOL

HISTORY OF BEAKERS, FLASKS, AND TEST TUBES

Michael Faraday believed that glassblowing was an important skill for chemists to have.

Hundreds of years ago, chemists used glassware they had in their homes for experiments. They used glass containers because glass is **transparent**. This allowed them to watch what happened during an experiment. In the 1800s, chemists such as Jöns Jacob Berzelius and Michael Faraday had the idea to design glassware specifically for experimenting. The test tube was created to test **chemical reactions**. Other chemists designed their own glassware to suit their needs. For example, John Joseph Griffin designed a short, wide beaker that is now called the Griffin beaker.

Chemical reactions often produce heat, and some chemicals can break glass. Chemists needed stronger containers. Otto Schott invented a very strong glass called borosilicate glass in 1893. This glass can withstand extremely high temperatures. It is still used today to make glassware for science labs.

Timeline of Lab Glassware

1827
Michael Faraday writes about his idea for a test tube in his book *Chemical Manipulation*.

1860
Emil Erlenmeyer, a German chemist, invents the Erlenmeyer flask.

1884
Otto Schott, Ernst Abbe, and Carl and Roderich Zeiss start the chemistry glassware company that will become SCHOTT AG.

2021
SCHOTT AG produces more than 1 billion borosilicate glass **vials** to store and deliver COVID-19 vaccines.

TYPES OF GLASSWARE

Beakers, flasks, and test tubes are the most common types of glassware used in a science lab. A beaker is shaped like a **cylinder** with a spout for pouring. Scientists can mix liquids in a beaker. Test tubes are also shaped like cylinders. They are taller and narrower than beakers. Flasks come in many shapes and sizes, and are used for different purposes.

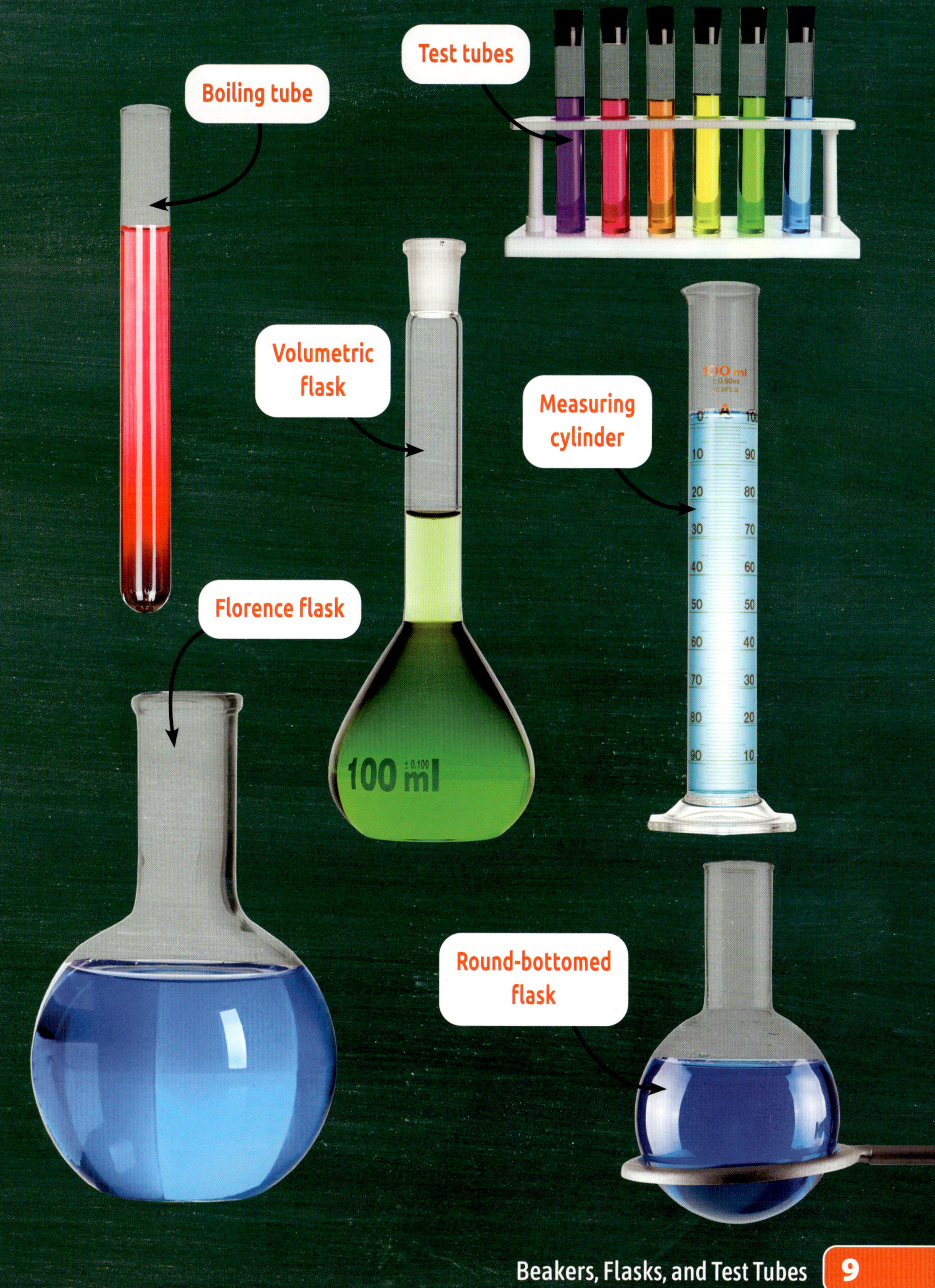
Test tubes
Boiling tube
Volumetric flask
Measuring cylinder
Florence flask
Round-bottomed flask
100 ml
100 ml

HOW THEY WORK

During a chemistry experiment, scientists may need to measure, mix, heat, cool, or store chemicals. Different glassware is used for each job. For example, a measuring cylinder can be used to measure the volume of a liquid more precisely than a beaker or flask.

Chemicals are often stored, transported, or mixed in beakers. A beaker's spout makes it easy to pour a liquid from the beaker to another container, such as a test tube or flask.

Chemicals react differently with each other. Some chemicals change color when they are mixed.

Test tubes are mostly used for mixing and heating small amounts of liquids. Boiling tubes, which are used to used to boil chemicals, have thicker glass and are a little wider than test tubes. They can tolerate very high temperatures.

Certain types of flasks are used in different experiments. The cone-shaped Erlenmeyer flask makes it easy to mix liquids without spilling them. A volumetric flask is used to measure an exact amount of liquid. Other flasks, such as the Florence and round-bottomed flasks, are often used for boiling liquids.

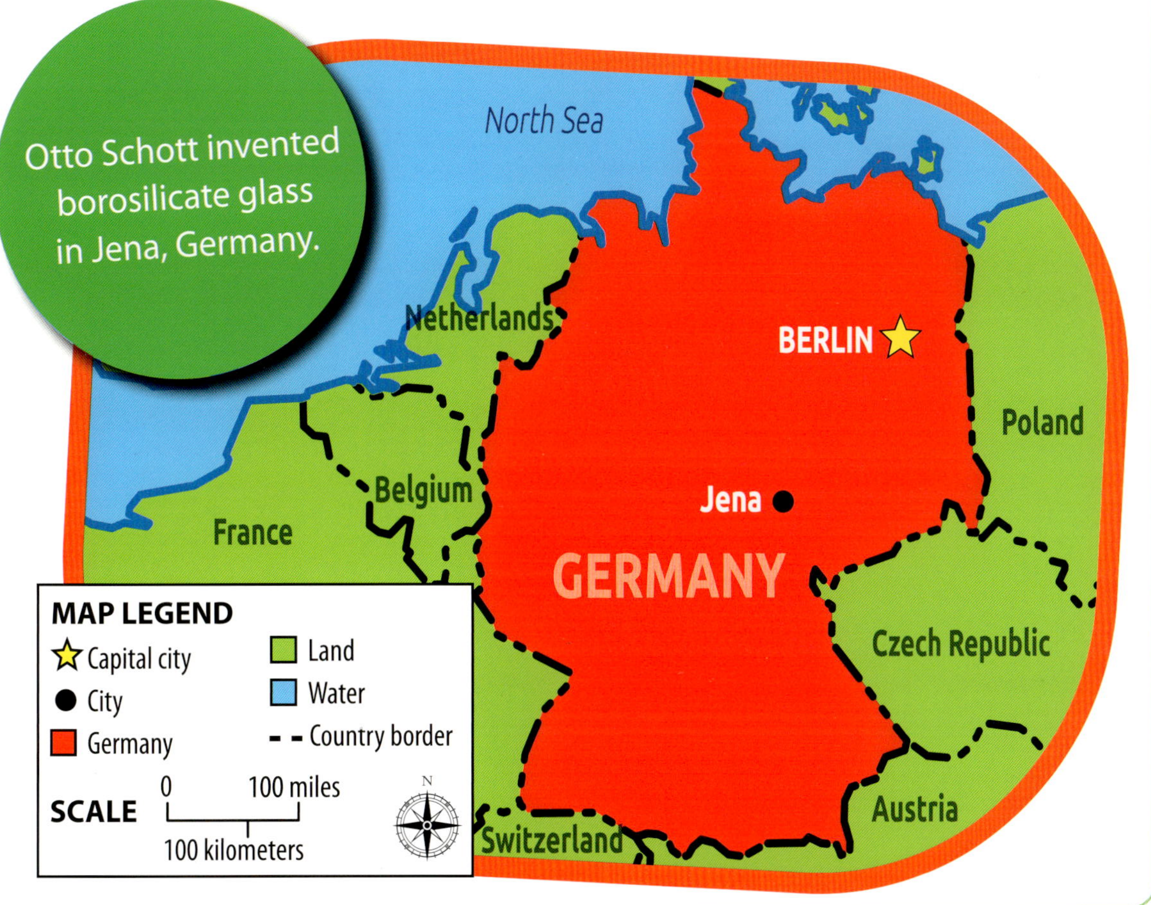

SAFETY IN THE LAB

Safety goggles, gloves, and a lab coat should be worn during experiments with dangerous chemicals.

Many chemicals can be dangerous on their own or when mixed with other chemicals. Never taste or directly sniff any chemical in the lab. It is important to understand the risks involved with each chemical. Some chemicals are highly flammable, which means they can easily catch fire.

Safety measures need to be taken around glassware, too. Remember, hot glass looks the same as cold glass. Do not pick up beakers, flasks, test tubes, or other glassware unless an adult says it is safe.

Glass can also break when it is dropped, or if it is heated or cooled too quickly. Avoid breaking glass containers by reading directions carefully. Check for cracks in glass before beginning an experiment. If glassware breaks, use **tongs** to pick up big pieces of broken glass. Use a dustpan and brush to clean up smaller pieces. Labs should have a special container for storing broken glass.

Since **1983**, all **dangerous chemicals** sold in the United States have had **warning labels** on the containers.

A colorless liquid called **carbon disulfide** is one of the **most flammable** chemicals used in science labs.

THE SCIENTIFIC METHOD

Scientists can discover new things by experimenting. A good science experiment can be repeated and get the same results. The scientific method helps scientists create experiments with reliable results.

1. QUESTION

People throughout history have wondered how and why certain things happen. Science helps answer these questions. What questions do you have?

2. RESEARCH

Scientists may have already asked the same questions as you. Start your experiment by researching the topic. What has already been discovered?

3. HYPOTHESIS

Use what you found out in your research to predict what will happen in your experiment. This is called a hypothesis.

4. EXPERIMENT

How can you find an answer to your question? Design an experiment that will test your hypothesis.

5. OBSERVATION

Watch closely to see what happens during your experiment. Draw or write your observations in a notebook.

6. ANALYSIS

Analyze your results. Consider what changed during the experiment. By doing the experiment, you caused something to happen. What happened? That is the result of your experiment.

7. CONCLUSION

Look at the results of your experiment. Does the **data** support the hypothesis? Tell others what you learned from your experiment.

EXPERIMENT

MYSTERY POWDER

Rosa likes to use chemistry to solve mysteries. During class one day, her teacher gives her an unknown white powder. Rosa must figure out what the white powder is. Her teacher tells her that it could be sugar, baking soda, or cornstarch. Can Rosa figure out what the white powder is with an experiment?

1. QUESTION What kind of white powder does Rosa have?

2. RESEARCH Rosa starts by looking closely at the powder. She has seen sugar, baking soda, and cornstarch at home. Based on its appearance, Rosa thinks the powder might be baking soda. She researches the **properties** of each powder and discovers that some solids, such as baking soda, will react with acids and create bubbles. Rosa learns that vinegar is a safe acid to experiment with.

SAFETY FIRST!

Even if you know what a solid or liquid is, it is not safe to eat or drink it. The container or your hands may have touched other chemicals that are dangerous.

3. HYPOTHESIS

The white powder is baking soda.

4. EXPERIMENT

Rosa puts one tablespoon of the white powder into a beaker. Then, she adds half a cup of vinegar to the beaker.

5. OBSERVATION

Rosa watches what happens as she pours in the vinegar. She observes bubbles forming and hears a hissing sound.

6. ANALYSIS

The powder reacted to the vinegar, causing a change. The result was that a gas was formed. The gas made bubbles form in the liquid.

7. CONCLUSION

Rosa's observations support her hypothesis. The white powder is baking soda, because sugar and cornstarch would not have reacted with the vinegar in that way.

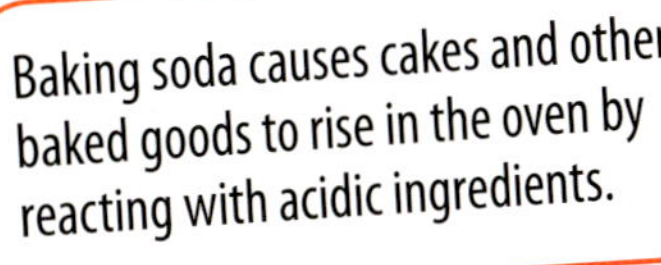

Baking soda causes cakes and other baked goods to rise in the oven by reacting with acidic ingredients.

EXPERIMENT

RUSTY NAILS

Have you ever noticed rust forming on a nail? Rust is created by a chemical reaction between iron and the water and **oxygen** in air. Kai wonders how rust forms and what might make it form faster. He notices that metal that gets wet often seems to have more rust than metal that usually remains dry. How can Kai learn more?

1. QUESTION

What makes metal rust faster than usual?

2. RESEARCH

Kai reads online that it is humidity, or water in the air, that causes rust. He wants to know if a nail will rust faster underwater than in air. Kai also learns that ocean water is salty. He wonders if the salt in the ocean makes the metal parts of boats and ships rust faster.

3. HYPOTHESIS

Both salt and water make metal rust faster.

4. EXPERIMENT

Kai sets up three test tubes in a test tube rack. He places one iron nail in each test tube. Kai fills the first test tube with plain water and the second test tube with salt water. He leaves the third test tube so the nail is only in air. After labeling the test tubes, he leaves them for three weeks.

SAFETY FIRST!

Broken glass is dangerous. When using glass equipment in an experiment, make sure it is stored in a safe place. If glass containers are left in the open, they may get knocked over and break.

5. OBSERVATION

When Kai returns to his test tubes, he notices rust forming on the first two nails. He sees that the nail in the second test tube, with the salt water, is much rustier than the first nail. The third nail, in the air, does not have any rust on it.

6. ANALYSIS

Water caused the iron nail in the first test tube to rust quickly. Rust formed because of a chemical reaction. Salt in the second test tube made the reaction occur even faster. The nail in the third test tube did not become rusty in the time provided.

7. CONCLUSION

The results of Kai's experiment support his hypothesis. Water causes iron to rust, and salt speeds up the reaction.

ACTIVITY

MAKE ELEPHANT TOOTHPASTE

Chemistry can be fun. Different liquids and solids may react to form bubbles or cause temperature changes. You can even use items in your house to make a foamy substance called elephant toothpaste. To be safe, research the chemicals you are using before mixing them together. Ask an adult for help with this experiment.

Materials

- Beaker or cup
- Erlenmeyer flask or a plastic bottle
- Packet of active yeast
- Warm water
- ½ cup hydrogen peroxide
- ¼ cup liquid dish soap

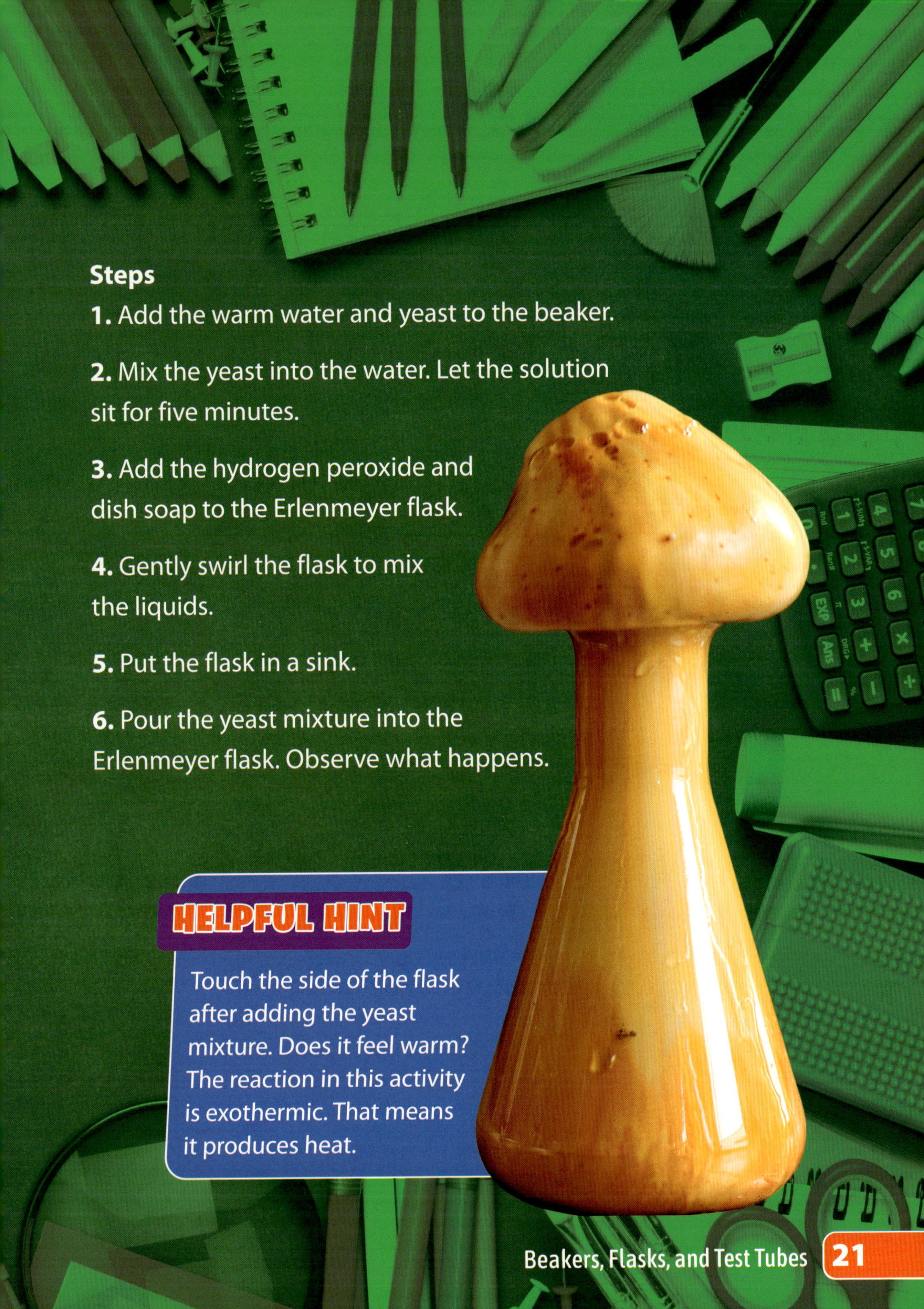

Steps

1. Add the warm water and yeast to the beaker.
2. Mix the yeast into the water. Let the solution sit for five minutes.
3. Add the hydrogen peroxide and dish soap to the Erlenmeyer flask.
4. Gently swirl the flask to mix the liquids.
5. Put the flask in a sink.
6. Pour the yeast mixture into the Erlenmeyer flask. Observe what happens.

HELPFUL HINT

Touch the side of the flask after adding the yeast mixture. Does it feel warm? The reaction in this activity is exothermic. That means it produces heat.

QUIZ

1 What is glassware?

2 What kind of glassware did Emil Erlenmeyer invent?

3 Is carbon disulfide flammable?

4 How are test tubes and boiling tubes different?

5 Who invented borosilicate glass?

6 What happens when vinegar and baking soda are mixed?

7 What does an exothermic reaction produce?

8 Where was borosilicate glass invented?

9 How does rust form?

10 How does baking soda cause cakes to rise?

Answers:
1. Glass containers, such as beakers, flasks, test tubes, and more **2.** The Erlenmeyer flask **3.** Yes **4.** Boiling tubes have thicker glass and are a little wider **5.** Otto Schott **6.** They react and form a gas **7.** Heat **8.** Jena, Germany **9.** Iron reacts with water and oxygen in the air **10.** By reacting with acidic ingredients

KEY WORDS

chemical reactions: when one or more substances change into one or more different substances

chemicals: any liquids, solids, or gases

cylinder: an object with a circular top and bottom that are the same size and are connected by a curved surface

data: facts or numbers collected for analysis

equipment: the tools used in a science experiment

experiments: scientific procedures performed to discover something

oxygen: a colorless gas that is needed for many chemical reactions

properties: specific physical and chemical qualities of a substance

tongs: a tool that can be used to pick up and hold items

transparent: easy to see through

vials: small glass containers

INDEX

Abbe, Ernst 7
alchemy 4

baking soda 16, 17, 22
beakers 4, 6, 8, 10, 13, 17, 20, 21, 22
Berzelius, Jöns Jacob 6
boiling tube 9, 11, 22
borosilicate glass 6, 7, 11, 22

carbon disulfide 13, 22
Chemical Manipulation 7
chemicals 4, 6, 7, 10, 11, 12, 13, 17, 18, 19, 20
chemistry 4, 7, 10, 16
cornstarch 16, 17
COVID-19 vaccines 7

dish soap 20, 21

Erlenmeyer, Emil 7, 22
experiments 4, 6, 10, 11, 12, 13, 14, 15, 16, 17, 18, 19, 20

Faraday, Michael 6, 7
flasks 4, 6, 7, 8, 9, 10, 11, 13, 20, 21, 22

glassware 4, 6, 7, 8, 10, 13, 22
gloves 12
Griffin, John Joseph 6

humidity 18
hydrogen peroxide 20, 21

iron 18, 19, 22

Jena, Germany 11, 22

laboratories 4, 6, 8, 12, 13
liquids 4, 8, 10, 11, 13, 17, 20, 21

measuring cylinder 9, 10

oxygen 18, 22

rust 18, 19, 22

safety 12, 13, 16, 17, 19, 20
safety goggles 12
salt 18, 19
SCHOTT AG 7
Schott, Otto 6, 7, 11, 22
solids 4, 16, 17, 20
sugar 16, 17

test tube 4, 6, 7, 8, 9, 10, 11, 13, 19, 22

vinegar 16, 17, 22

yeast 20, 21

Zeiss, Carl 7
Zeiss, Roderich 7

LIGHTBOX

SUPPLEMENTARY RESOURCES

Click on the plus icon ⊕ found in the bottom left corner of each spread to open additional teacher resources.

- Download and print the book's quizzes and activities
- Access curriculum correlations
- Explore additional web applications that enhance the Lightbox experience

LIGHTBOX DIGITAL TITLES

Packed full of integrated media

VIDEOS

INTERACTIVE MAPS

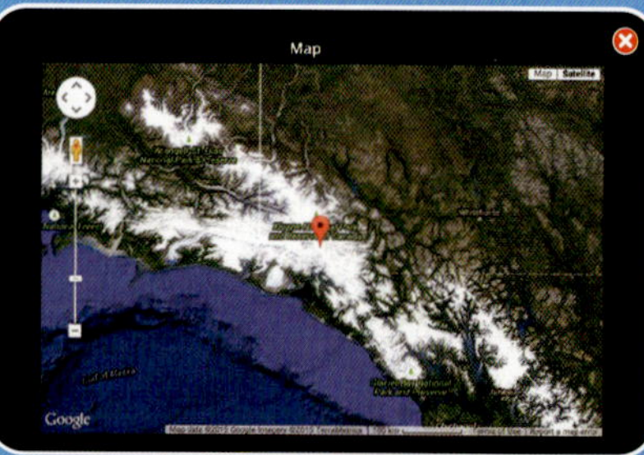

WEBLINKS

SLIDESHOWS

A cirque is a rounded, bowl-shaped area where snow collects

QUIZZES

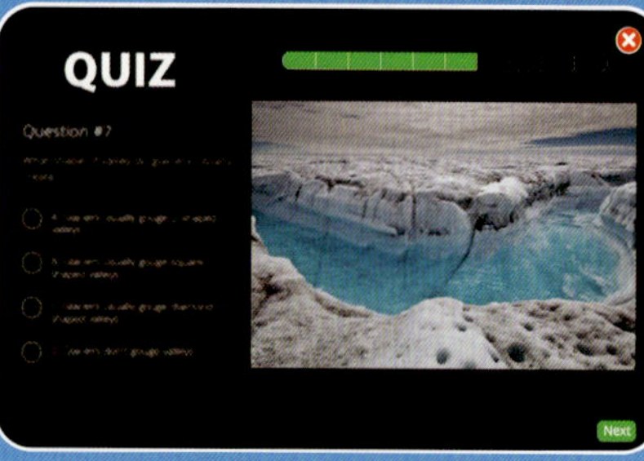

OPTIMIZED FOR

- ✓ TABLETS
- ✓ WHITEBOARDS
- ✓ COMPUTERS
- ✓ AND MUCH MORE!

Published by Lightbox Learning
276 5th Avenue, Suite 704 #917
New York, NY 10001
Website: www.openlightbox.com

Library of Congress Cataloging-in-Publication Data

Names: Noelle, Becky, author.
Title: Beakers, flasks, and test tubes / Becky Noelle.
Description: New York, NY : Smartbook Media Inc., [2022] | Series: Science lab equipment and safety | Includes index. | Audience: Grades: 2-3
Identifiers: LCCN 2021032117 (print) | LCCN 2021032118 (ebook) | ISBN 9781510559028 (library binding) | ISBN 9781510559035
Subjects: LCSH: Scientific apparatus and instruments--Juvenile literature. | Science--Experiments--Juvenile literature.
Classification: LCC Q185.3 .N64 2022 (print) | LCC Q185.3 (ebook) | DDC 502.8--dc23
LC record available at https://lccn.loc.gov/2021032117
LC ebook record available at https://lccn.loc.gov/2021032118

Printed in Guangzhou, China
1 2 3 4 5 6 7 8 9 0 25 24 23 22 21

092021
111020

Project Coordinator: Priyanka Das **Designer:** Ana María Vidal

Every reasonable effort has been made to trace ownership and to obtain permission to reprint copyright material. The publisher would be pleased to have any errors or omissions brought to its attention so that they may be corrected in subsequent printings. The publisher acknowledges Alamy, Getty Images, and Shutterstock as its primary image suppliers for this title.